Math Practice Simplified

Tables & Graphs

Author
Sharon Schwartz

Cover Design
Elliot Kreloff, Inc.

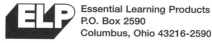

Essential Learning Products
P.O. Box 2590
Columbus, Ohio 43216-2590

Copyright ©2001 Essential Learning Products

Permission to Reproduce
Permission to reproduce this material is granted for classroom and noncommercial use only. Reproduction for an entire school or school system is prohibited. No part of this publication may be stored in a retrieval system or transmitted in any form by any means without the publisher's permission.

The text in this book originally appeared in *Math Practice Book—Tables, Graphs, Statistics, Probability*
©1991 Essential Learning Products

Printed in the United States of America
10 9 8 7 6 5 4 3 2

INTRODUCTION

Reading and interpreting information from tables, charts, and graphs is an essential life skill. *Math Practice Simplified—Tables & Graphs* contains high-interest, realistic activities that help students understand the importance of this skill. Some work with statistics and probability is also provided.

Math Practice Simplified—Tables & Graphs is exactly what the title says. It is practice material presented in its most direct and simplified form. Exercises are presented in student-friendly layouts with few distractions to interfere with concentration.

It is best if computation operations are mastered *before* a student proceeds to the concepts presented here. The brief explanations/examples provided throughout the book are there to serve as reminders; they are *not* meant to be instructive. Teaching of a given concept must be done before this book is used.

As with any skill, teaching and understanding of concepts and processes must precede practice. And as important as that understanding is, no skill is truly accomplished without practice. Individuals require varying amounts of practice to acquire facility. The use of this book can be adapted to serve each child's need. It can be used to reinforce skills one page at a time or for a specified amount of time. However, studies have proven that *brief* daily practice is generally more effective than longer but less frequent practice. Having a child work for more than thirty minutes at a time is more apt to be discouraging than helpful.

In this book, students practice reading a variety of tables and graphs. They then construct their own tables and graphs from a set of data, which they use to solve problems. Students are also given the opportunity to gather data and create graphs of their own in both prescribed and open-ended situations. These lessons can be great collaborative activities. Numbers and computations have been kept simple. Emphasis is on using graphs to make decisions, estimations, and predictions.

Sometimes graphs can be misleading, so students are shown how to determine whether a particular graph gives a false impression.

Graphs and tables are often used in conjunction with statistics. Lessons on mean, median, mode, and range are presented. To interpret large amounts of statistical data at a glance, students become familiar with reading and making scattergrams, stem and leaf plots, line plots, box plots, histograms, and frequency polygons.

A few pages are devoted to helping students understand probability. Independent and dependent events are included, as are lessons on tree diagrams and combinations and permutations. Students are also able to observe ways in which probability can be used in experiments to predict outcomes.

In most cases, there is not room for students to work directly on the page. Allow students to work on scrap paper or to work neatly on a separate sheet so their work can be checked.

Answers are provided at the back of the book.

This book is designed for students in grades 5-8 and for older students who may need additional practice. Students using *Math Practice Simplified—Tables & Graphs* have the opportunity to build a solid foundation for mathematics, increase self-esteem upon successful completion, and improve performance on standardized tests.

CONTENTS

Pages

Tables
- 4-7 Reading and Using Tables
- 8-11 Making and Using Tables
- 12-18 Making and Using Tables/Consumer Economics

Graphs
- 19-22 Bar Graphs
- 23 Double-Bar Graphs
- 24 Bar Displays/Divided-Bar Graphs
- 25-28 Pictographs
- 29-32 Circle Graphs
- 33-36 Line Graphs
- 37-38 Double-Line Graphs
- 39-41 Selecting a Graph to Display Data
- 42 Using Graphs to Estimate
- 43 Using Graphs to Predict
- 44-45 Evaluating Graphs

Pages

Statistics
- 46 Mean, Median, Mode, Range
- 47 Scattergrams
- 48 Stem and Leaf Plots
- 49 Line Plots/Box Plots
- 50-51 Histograms/Frequency Polygons

Probability
- 52 Experimental Probability
- 53 Theoretical Probability
- 54 Independent Events
- 55 Dependent Events
- 56 Odds
- 57 Tree Diagrams
- 58 Combinations and Permutations
- 59 Experimental/Actual Probability
- 60-64 **Answers**

**NCTM Standards for *Math Practice Simplified*
Tables and Graphs**

Concepts in this book are based on *Principles and Standards for School Mathematics* as identified by the National Council of Teachers of Mathematics. Lessons correlate to the following specific standards:

Problem Solving
- Apply and adapt a variety of appropriate strategies to solve problems.
- Build new mathematical knowledge through problem solving.
- Solve problems that arise in mathematics and in other contexts.

Data Analysis and Probability
- Represent data using tables and graphs such as line plots, bar graphs, and line graphs.
- Describe parts of the data and the set of data as a whole to determine what the data show.

Write a few sentences about what you already know about tables and graphs.

_____
_____
_____
_____
_____
_____
_____
_____

Did you know that a class schedule is a kind of table? Choose a day of the week and show your schedule.

Tables show information in an organized way to help you find what you need quickly.

Amount of Television Watched by People Age 2-17

hours per day	number of people
Less than 1	1,010,000
2	12,760,000
3	25,840,000
4	13,910,000
5 or more	1,540,000

- The table's title tells you what the information is about.
- The headings show what information is listed.
- The information is organized for you in the body of the table.

Example: To find how many people age 2-17 watch television 4 hours per day, look down the column *hours per day* until you find 4. Then look across to the right. You find that 13,910,000 people age 2-17 watch television 4 hours per day.

Use the table to answer the questions.

1. How many entries are listed under *hours per day*? _____

2. How many hours per day do most people under 18 watch television? _____

3. How many hours per day do the least number of people under age 18 watch television? _____

4. How many watch 5 or more hours of television per day? _____

5. How many watch 3 or fewer hours of television per day? _____

6. How many people age 2-17 were surveyed? _____

Reading and using tables

Here's the menu and a tax table from the Star Diner.

Star Diner Menu

Specials
Hamburger $3.25
Grilled Chicken $3.95
Spaghetti $3.50

Fish $4.75
Veggie Burger $3.35

Beverages
Cola, Orange, Uncola
Sm $0.75....................Lrg $1.25
Coffee $0.95
Tea $0.55
Milk $0.70
Spring or Mineral Water ..$0.90

Side Orders
Soup of the Day $1.75
French Fries $1.75
Baked Potato $1.50
Cole Slaw $0.95
Garden Salad $2.15
Tossed Salad $2.75
Garlic Bread $1.50
Fruit Plate $2.50

Tax Table 6% Sales Tax

Amount	Tax
$7.75–7.91	$0.47
$7.92–8.08	$0.48
$8.09–8.24	$0.49
$8.25–8.41	$0.50
$8.42–8.58	$0.51
$8.59–8.74	$0.52
$8.75–8.91	$0.53
$8.92–9.07	$0.54
$9.08–9.24	$0.55
$9.25–9.40	$0.56
$9.41–9.58	$0.57

Example: To find the tax on $8.79, look in the tax table amount column for the entry that includes $8.79. It is $8.75-$8.91. The tax on $8.79 is $0.53.

Use the information from the menu and tax table to complete these bills.

1. Guest Check
Hamburger _____
French Fries _____
Garden Salad _____
Large Cola _____
Subtotal _____
Tax _____
Total _____

2. Guest Check
Veggie Burger _____
Tossed Salad _____
Fruit Plate _____
Mineral Water _____
Subtotal _____
Tax _____
Total _____

3. Guest Check
Fish _____
Baked Potato _____
Cole Slaw _____
Tea _____
Subtotal _____
Tax _____
Total _____

4. Guest Check
Soup _____
Grilled Chicken _____
Garlic Bread _____
Coffee _____
Subtotal _____
Tax _____
Total _____

ROMANO'S PIZZA

	Small	Medium	Large
Cheese:	$7.50	$8.00	$8.50
+ 1 topping	$9.00	$9.50	$10.00
+ 2 toppings	$10.50	$11.00	$11.50
Combination: cheese, pepperoni, mushrooms, sausage	$11.50	$12.20	$13.50
Supreme Combination: green peppers, olives, tomatoes, and extra cheese	$13.75	$14.50	$16.50

Extra Cheese add $1.50   Tax included on all items.

Example: To find the cost of a small pizza with mushrooms and sausage, look over from the *+ 2 toppings* row and down from the "small" column. The pizza will cost $10.50.

Use the menu to answer these questions.

1. What is the cost of a large cheese pizza with peppers? _____

2. How many medium cheese pizzas with mushrooms and pepperoni can you buy with $30? _____ How much change will you get? _____

3. Chris ordered a large cheese pizza and a small cheese pizza with pepperoni. Will $20 be enough to pay for the pizzas? _____

4. Karen and Bill split the cost of a medium combination pizza with extra cheese. How much did each pay? _____

5. Susan and three friends will split the cost of a large cheese pizza with mushrooms and extra cheese. About how much will each pay? _____

6. Cory and two friends will split the cost of a large combination pizza. How much change will each of them receive from $15? _____

7. Melinda is having a pizza party for twenty people, including herself. There are eight slices in a large pie. She figures each person will eat 2 slices. How much will the pizzas cost if they are all large supreme pizzas? _____

Complete the table.

Chocolate Factory Payroll for Week of April 14

Name	Hourly Rate	Hours Worked					Total Hours	Gross Pay
		M	T	W	TH	F		
1. Emin Emms	$5.25	5	4	4	5	4		
2. Candy Barr	$6.75	6	3	4	5	7		
3. Reesy Kupp	$5.50	6	6	6	6	6		
4. Hirshy Khiss	$6.00	7	7	7	6	5		
5. Kit Katt	$6.10	8	7	6	8	7		
6. Carl Mello	$6.35	8	8	5	6	8		
7. Elmund Joy	$5.85	7	6	8	8	8		

Use information from the Hours Worked table below to complete the Tally Chart. Begin by putting a mark in the tally column for each number of hours worked per day. Cross off each number after you tally it. (The first three have been done.) Then count the tally marks and write the number in the frequency column.

Hours Worked per Day

5̶	4̶	4̶	5	4	6	3	4	5
7	6	6	6	6	6	7	7	7
6	5	8	7	6	8	7	8	8
5	6	8	7	6	8	8	8	

Tally Chart
Number of hours worked per day

number of hours	tally	frequency
3		
4	||	
5	|	
6		
7		
8		

1. Which number of hours worked has a frequency of 4? _____

2. What is the most common number of hours worked per day? _____

   the least common? _____

Making and using tables

The Chaps and Britches riding club went on a ride-a-thon for handicapped riders. The table shows how many miles each rider rode.

Person	Marta	Dory	Mark	Meg	Dave	Kim	Joe	Len	Jeff	Carol
miles	20	11	7	13	15	8	5	9	17	12

This table shows how much money was pledged by companies for each mile ridden.

Company	Computer Net	Mick E. Moose Co.	Bubble Gum Unlimited	Lemon Used Cars	Video Vault
amount pledged per mile	$0.20	$0.10	$0.25	$0.50	$1.00

Use the information from the tables to answer these questions.

1. How far did Dory ride? _____

2. How much did Video Vault pledge per mile? _____

3. How much did Video Vault pay for the miles Dory rode? _____

4. How much did Bubble Gum Unlimited pay for the miles Jeff rode? _____

5. How much did Computer Net pay for the miles Mark rode? _____

6. How much did Lemon Used Cars pay for the miles Carol rode? _____

7. How much per mile was pledged in all? _____

8. What was the total amount collected for the miles Marta rode? _____

9. How many miles were ridden in all? _____

10. What was the total amount of money brought in by the ride-a-thon? _____

Timetables for most trains and buses show departure times. Here is the timetable for two trains.

R&R Railroad Timetable

Station	Green Line Local		The Rocket Express
Becker	Departs	7:00 a.m.	8:00 a.m.
Cattle City	↓	7:18 a.m.	↓
Treasure Cove	↓	7:51 a.m.	↓
River Falls	↓	8:07 a.m.	↓
Kingston	Arrives	8:40 a.m.	9:00 a.m.

1. About how long does it take the Green Line Local to go from Cattle City to River Falls? _____ From Treasure Cove to Kingston? _____

2. How much faster is it to take The Rocket Express from Becker to Kingston? _____

3. If it is 80 miles from Becker to Kingston, what is the average speed of the Green Line Local? _____ The Rocket Express? _____

R&R Railroad Fares

	Becker	Cattle City	Treasure Cove	River Falls	Kingston
Becker		$2.00	$4.20	$4.50	$8.00
Cattle City	$2.00		$2.20	$2.40	$4.90
Treasure Cove	$4.20	$2.20		$ .40	$2.75
River Falls	$4.50	$2.40	$ .40		$2.50
Kingston	$8.00	$4.90	$2.75	$2.50	

4. Use the fare schedule to find out how much it costs to go from

Becker to Kingston _____   Cattle City to Treasure Cove _____

River Falls to Becker _____   Kingston to Treasure Cove _____

Using transportation tables

Work with a partner. Each of you choose the place in the United States that you would most like to visit.

Try to find out the cost and travel schedule if you take a bus, a train, or a plane. Remember, fares often vary according to when you travel. Show a chart or table of your findings.

Write a paragraph about which mode of travel you would choose and tell why.

On Your Own—Tables

Collision insurance pays to replace or repair damage to a car caused by an accident.

Comprehensive insurance pays for repair or replacement of a car due to fire, theft, vandalism, or an act of nature.

Liability insurance protects you if someone is injured or property is damaged in an accident.

Deductible (ded) means the amount you pay for damage before the insurance company begins to pay.

The total annual **premium** is the sum of the amounts for the types of coverage purchased.

\multicolumn{2}{c}{}	\multicolumn{6}{c}{**Annual Auto Insurance Premiums** (in dollars)}
		\multicolumn{2}{c}{Collision}	\multicolumn{2}{c}{Comprehensive}	\multicolumn{2}{c}{Liability}
car	driver	$100 ded	$500 ded	no ded	$50 ded	$300,000	$500,000
Class B	Teen	884.40	645.70	291.30	253.50	357.00	382.40
	Adult	271.00	197.90	92.20	80.30	110.80	118.60
Class C	Teen	1098.00	801.70	691.80	601.80	357.00	382.40
	Adult	336.50	245.70	219.10	190.60	110.80	118.60

Use the table to find these answers.

1. What is the annual premium for collision insurance with $100 deductible for a teen with a Class C car? _____

2. What is the annual premium for liability insurance for an adult with coverage for $500,000? _____

3. What is the total annual premium for a teen with a Class B car for collision insurance with $100 deductible, comprehensive with no deductible, and liability coverage of $300,000? _____

4. What is the total premium for an adult with a Class C car for collision insurance with $100 deductible, comprehensive with $50 deductible, and liability of $300,000? _____

5. What is the total premium for an adult with a Class B car for collision insurance with $500 deductible, comprehensive with $50 deductible, and liability coverage of $500,000? _____

A hair dryer uses 1600 watts of electricity an hour. A stereo uses 150 watts, a refrigerator 750 watts, an electric stove 12,000 watts, a color TV 200 watts, a computer 300 watts, and a 100-watt light bulb 100 watts. Make a chart to show information about the amount of electricity used by these appliances for 1 hour, 2 hours, 3 hours, 6 hours, 8 hours, and 24 hours.

Appliance	1 hr.	2 hr.	3 hr.	6 hr.	8 hr.	24 hr.

Suppose it costs 8¢ to use 1000 watts. Use the information from the table you made to answer the questions. Express all answers to the nearest cent.

1. How much does it cost to run a color TV for 8 hours? _____

2. How much does it cost to run a refrigerator for 24 hours? _____

3. How much does it cost to run an electric stove for 3 hours? _____

4. How much does it cost for a 100-watt bulb to be lit for 6 hours? _____

5. How much does it cost to operate a computer for 2 hours? _____

6. Does it cost more or less than a penny for a stereo to play for a half hour? _____

7. How much does it cost to run a hair dryer for 15 minutes? _____

Making and using tables

To pay back a loan, you usually pay the amount borrowed plus a finance charge (interest) in monthly installments. The amount you pay monthly depends on the amount borrowed, the time you take to repay, and the **A**nnual **P**ercentage **R**ate (APR).

Monthly Payments for Amount Borrowed

APR	time	$100	$500	$1000	$10,000
7%	12 mo.	$8.65	$43.27	$86.54	$856.36
	24 mo.	$4.48	$22.39	$44.78	$447.78
	36 mo.	$3.09	$15.44	$30.88	$308.81
	60 mo.	$1.98	$9.90	$19.80	$198.03
12%	12 mo.	$8.89	$44.43	$88.87	$888.66
	24 mo.	$4.71	$23.54	$47.08	$470.82
	36 mo.	$3.32	$16.51	$33.22	$332.21
	60 mo.	$2.22	$11.12	$22.25	$222.49

Example: To find the monthly payment for $500 borrowed for 24 months at an APR of 7%, look in the 7% APR block, find 24 months, and look across to $500. Monthly payments will be $22.39.

Use the table to solve.

1. Megan had saved some money to buy a bicycle. However, she needed to borrow $100 for 12 months at an APR of 7%.

   What are her monthly payments? _____

   Find the amount she will repay. _____
   (Multiply the monthly payment by the number of payments.)

2. Greg borrowed $500 for 24 months at an APR of 12% to buy a used car.

   What are his monthly payments? _____

   How much will he repay? _____

3. Stephanie borrowed $1000 for 36 months at 12% to buy a computer.

   How much will she repay? _____

   Find the total finance charge. _____
   (Subtract the amount borrowed from the amount repaid.)

4. Brian borrowed $10,000 at 7% for 60 months to buy a boat.

   What is the finance charge? _____

When buying merchandise from a catalog or online, shipping and handling charts are often provided so that you can compute these charges.

A. AM/FM Radio   Wt. 3.1 lb.
416-5342 $29.99

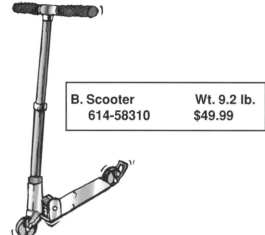

B. Scooter   Wt. 9.2 lb.
614-58310 $49.99

Shipping and Handling Charges			
Pounds	Zone 1	Zone 2	Zone 3
6–6.9	$3.92	$4.16	$4.56
7–7.9	$4.04	$4.26	$4.72
8–8.9	$4.16	$4.36	$4.80
9–9.9	$4.28	$4.46	$4.88
10–10.9	$4.40	$4.58	$4.96
11–11.9	$4.52	$4.70	$5.06
12–12.9	$4.64	$4.82	$5.19
13–13.9	$4.73	$4.94	$5.29
14–14.9	$4.82	$5.09	$5.42
15–15.9	$4.91	$5.24	$5.55
16–16.9	$5.00	$5.39	$5.77
17–17.9	$5.09	$5.57	$5.99

Example: To find charges for a 7.1-lb. item that is being shipped to Zone 2, look under the pounds column for the interval that includes 7.1. Then look across that row and down from the Zone 2 column.
The shipping and handling charge is $4.26.

Use the information from the product descriptions above and the shipping chart to complete this order form. The package is going to Zone 3.

Item	Catalog Number	How many	Price for 1	Total Price	Shipping Weight lb
Radio		2			
Scooter		1			
			Total		
			Tax	$6.59	
			Shipping and Handling		
			Total Cost		

Completing order forms

Find a catalog and pretend to order at least five items. Attach the order form here.

If a computer is available, you might want to work with an adult and log on to a Web site that sells books, toys, music, or one of your other favorite things. Practice completing the order form with at least five items and, if possible, print out your results to attach to this page. Be careful not to actually enter the order unless that is your intention and the adult agrees.

Write a paragraph about the process.

_____
_____
_____
_____

What was the easiest part?

_____
_____
_____

What was not easy?

_____
_____
_____

List any concerns you might have.

_____
_____
_____
_____

On Your Own—Order form

This is a sample section of a stock table that can be found in the financial section of a newspaper.

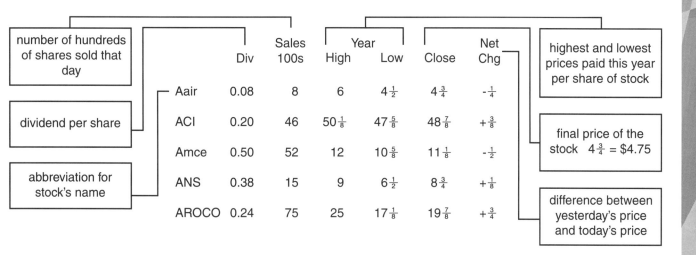

Example: The cost of 200 shares of Alexis Air stock at closing today is 200 × $4.75, or $950.
The net change of $-\frac{1}{4}$ of a dollar means yesterday the price was $0.25 more per share.

Use the stock chart to answer the questions.

$\frac{1}{8}$ of a dollar = $0.125

1. What is the dividend paid on Amce stock? _____

2. How many shares of Aair stock were sold today? _____

3. What was this year's high for ANS? _____

4. What was the closing price of ACI today? _____

5. Michael bought 300 shares of Aair at today's closing price.

   How much did he pay for the share of stock? _____

6. Dana bought 400 shares of AROCO stock at yesterday's closing price.

   How much did she pay for the stock? _____

   How much more is the stock worth today? _____

Reading and using tables

Work with a partner. Find a recent copy of a stock table. Take turns writing and answering questions about three different stocks. Show your work below.

Bar graphs make it easy to compare information quickly.
To read a bar graph, find the bar that represents the information you want. Look from the edge of the bar in a straight line to the scale. Read the value from the scale.

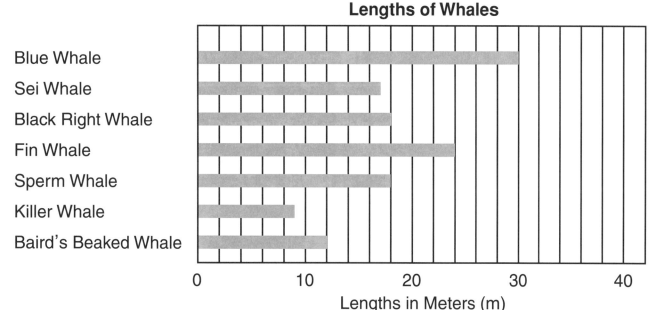

Example: The blue whale is 30 meters long.

Use the bar graph to answer the questions.

1. What is the bar graph about? _____

2. Which whale on the graph is the longest? _____

3. Which whale on the graph is the shortest? _____

4. Which whale is longer, the fin whale or the sperm whale? _____

5. What distance does each mark on the scale represent? _____

6. What is the difference between the length of the Baird's beaked whale and the length of the killer whale? _____

7. What is the difference between the length of the longest whale and the length of the shortest whale? _____

Reading and using bar graphs

Sometimes it's easier to get information from a graph. Other times it's easier to get the information from a table.

Height of Some of the Tallest Skyscrapers in the World

Building	Location	Height	Stories
Petronas Towers	Kuala Lumpur, Malaysia	1483 ft.	88
Sears Tower	Chicago, IL, USA	1450 ft.	110
Jin Mao Tower	Shanghai, China	1380 ft.	88
Empire State Building	New York, NY, USA	1250 ft.	102

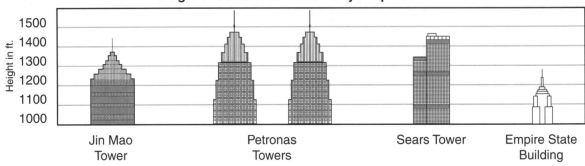

Answer the questions. Check which was easier to use.

	Table	Graph	Either
1. How tall is the Jin Mao Tower? _____	☐	☐	☐
2. Which is taller, the Empire State Building or the Sears Tower? _____	☐	☐	☐
3. Where are the Petronas Towers located? _____	☐	☐	☐
4. Which building is tallest? _____	☐	☐	☐
5. What is the difference between the heights of the tallest and shortest buildings? _____	☐	☐	☐
6. How many buildings are taller than 1300 feet? _____	☐	☐	☐

Evaluating use of graphs vs. tables

Here are the steps for making a bar graph.

Step 1: Decide whether to make a vertical or horizontal graph.
(Will the necessary number of bars fit better horizontally or vertically?)

Step 2: Determine a scale. (Find the largest value. Determine a measurement to be equal to 1 unit. Will the longest bar fit? If not, try 2 units for each and so on until you find a scale that will fit the space for the graph.)

Step 3: Draw the left and bottom sides of the graph. Mark the scale and label the units represented. Decide how wide to make each bar and how much space to leave between. Mark the widths of the bars and label them. Draw the other two sides.

Step 4: Use the scale to draw the bars.

Step 5: Give the graph a title that tells what the information is about.

Make a bar graph for the information in the table.

maximum speeds in miles per hour							
Cheetah	70	Human	27.9	Lion	50	Zebra	40
Reindeer	32	Grizzly Bear	30	Squirrel	12	Chicken	9

1. Can you catch a lion? _____

2. Can a reindeer outrun a grizzly bear? _____

Making and using bar graphs

Work with a partner. Gather some information about outer space, space travel, or planets.

Write a paragraph describing the research process you used.

_____
_____
_____
_____
_____
_____
_____
_____

Create a bar graph to show the results of your findings.

On Your Own—Bar graph

A **double-bar graph** shows two sets of related data on one graph. This graph shows the favorite types of music of a group of seventh graders.

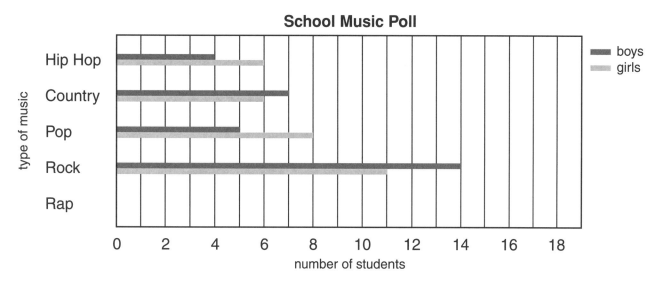

1. Use this information to complete the graph:
   9 boys and 6 girls chose rap music as their favorite.

Use the double-bar graph to answer the questions.

2. How many boys chose pop as their favorite music? _____

3. How many girls chose hip hop as their favorite music? _____

4. Did more boys or girls choose rock music as their favorite? _____

5. For what type of music is the difference between girls' and boys' choice the least? _____

6. Arrange the choices of types of music for girls and boys in order from most favorite to least favorite.

   Girls: _____

   Boys: _____

7. How many students were surveyed? _____

Reading and using double-bar graphs

Data can be compared by the lengths of bars on a display. Create a bar display to show the information in this article.

> The Chunnel, the 31-mile railway tunnel connecting France and Britain, is one of the world's longest railway tunnels. The Seikan (33.5 mi.), Dai-shimizu (14 mi.), and the Kanmon (12 mi.), are all located in Japan. The Apennine, in Italy, is 11 miles long. The Simplon I and II, which connects Switzerland and Italy, is 12 miles long.

A **divided-bar graph** shows the relationship of information on one bar. This divided-bar graph shows the results of a survey at a toy store when purchasers were asked: "Who will get the toys you purchased?"

| Sons 35% | Daughters 25% | Grandchildren 25% | Other 10% | Self 5% |

1. Which group gets the most toys? _____

2. Make a divided-bar graph that shows the time, over a 24-hour period, that you spend sleeping, eating, at school, at other outdoor activities, and at other indoor activities. (Show times to the nearest hour.)

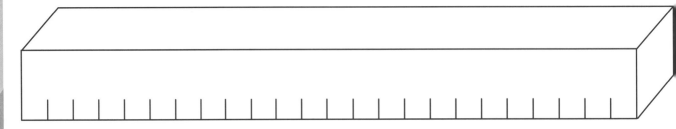

Creating and using bar graphs and divided-bar graphs

Pictographs are picture graphs. To read a pictograph, find the scale that shows how much or how many each picture represents.

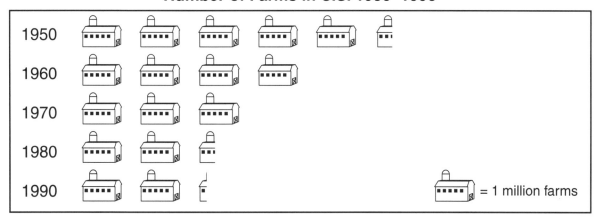

Example: The scale shows that each barn represents 1 million farms. To find the number of farms in 1990, count the number of barns pictured—about $2\frac{1}{5}$. Then multiply 1,000,000 by that number. $1,000,000 \times 2\frac{1}{5} = 2,200,000$. There were 2,200,000 farms in 1990.

Use the pictograph to answer the questions.

1. About how many farms were there

   a. in 1950 _____ b. in 1960 _____

   c. in 1970 _____ d. in 1980 _____

2. How many more farms were there in 1960 than 1970? _____

3. When was the number of farms more than double the amount in 1980? _____

4. How has the number of farms changed from 1950 to 1990? _____

5. How would you represent 1,500,000 farms? _____

6. Suppose each barn represented 500,000 farms. How many should be drawn for 1980? _____

Reading and using pictographs

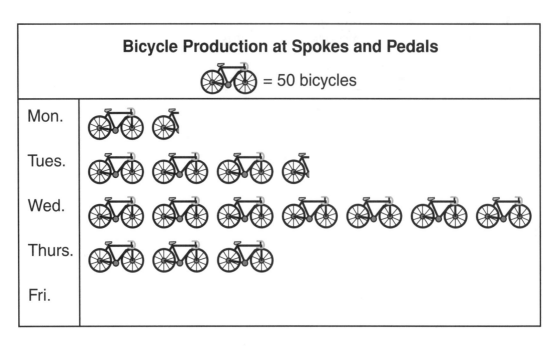

Use the pictograph to answer the questions.

1. Spokes and Pedals, Inc. made 200 bicycles on Friday. Look at the pictograph above. How many bicycles are needed to complete the graph?

2. How might you check to see that you drew the correct number of bicycles?
   _____

3. What other way might you display the data?
   _____

4. What does 🚲 mean?
   _____

5. On what day are the most bicycles produced? _____

6. How many more bicycles were produced on Wednesday than Tuesday?
   _____

7. Suppose each bicycle represents 100 bicycles. How many bicycles would you draw for Wednesday? _____ For Monday? _____

8. How many bicycles were produced this week? _____

Creating and using pictographs

Here are the steps for making a pictograph.

Step 1: Decide on a picture you will use to represent the data. Determine a scale. (Let 1 picture = 1 unit. Will the number of pictures you need fit? If not, try 1 picture for 2 units and so on until you find a scale that will fit the space for the graph.)

Step 2: Draw the sides of the graph. Label the scale and the items to be represented. Mark spaces for the pictures with equal spaces in between.

Step 3: Use the scale to draw the correct number of pictures for each item.

Step 4: Give the graph a title that tells what the information is about.

Make a pictograph to show this data.

U.S. Population in Millions

year	1950	1960	1970	1980	1990	2000
number of people	152.3	180.7	205.1	227.2	249.5	281.4
	150					

Round to the nearest ten million.
The first one has been done for you.

Work with a partner. Together create a pictograph about pets—anything at all about pets.

First you must collect your data by conducting a survey or researching information on the Web or at the library. Write about what you did.

_____

_____

_____

_____

_____

_____

_____

Create your graph below.

On Your Own—Pictograph

Circle graphs show how parts of something compare to the whole.
This circle graph shows how each part of the Pocins' family's income is spent.

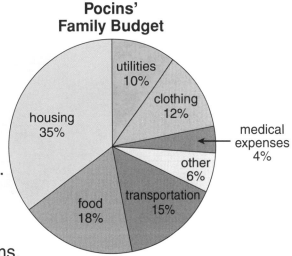

Pocins' Family Budget

Example: To find what percent of the family's income is used for transportation, locate the section labeled *transportation*. It represents 15% of the budget.

You can use the information to solve problems.
Suppose the family has a monthly take-home pay of $3000.
To find the amount budgeted for transportation, find 15% of $3000.

(15% is the same as $\frac{15}{100}$ or 0.15.)

0.15 x $3000 = $450 $450 is budgeted for transportation.

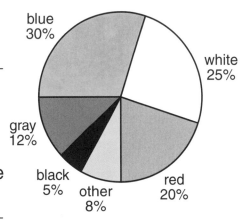

Favorite Car Colors

Use the graph at right to answer the questions.

1. Which color did most people prefer? _____

2. What color was chosen four times as often as black? _____

3. Which two colors combined did exactly half the group choose? _____

4. What percentage chose a color such as pink as their favorite color? _____

5. What *fraction* of the group chose white as the favorite color? _____

6. If 200 people were surveyed, how many chose gray? _____

7. If 450 people were surveyed, how many chose red? _____

Reading and using circle graphs

The Bookworms kept this tally chart of sales at a book fair.

Tally	
Mysteries	𝍷𝍷𝍷𝍷 𝍷𝍷𝍷𝍷 𝍷𝍷𝍷𝍷 𝍷𝍷𝍷𝍷 𝍷𝍷𝍷𝍷 𝍷𝍷𝍷𝍷 𝍷𝍷𝍷𝍷 𝍷𝍷𝍷𝍷
Novels	𝍷𝍷𝍷𝍷 𝍷𝍷𝍷𝍷 𝍷𝍷𝍷𝍷 𝍷𝍷𝍷𝍷 𝍷𝍷𝍷𝍷 |||
Reference	𝍷𝍷𝍷𝍷 |||
Other	||||

Use information from the tally chart above to complete this table. Part of the first one is done for you.

Bookworms' Book Sale

Total Books Sold: _____			
Type of Book	Number Sold	Fraction of all Books Sold	Percent Sold
Mysteries	☐	$\frac{\Box}{80} = \frac{1}{2}$	$\frac{1}{2} = \frac{50}{100} = 50\%$
Novels			
Reference			
Other			

Use the information from the table to complete the circle graph.

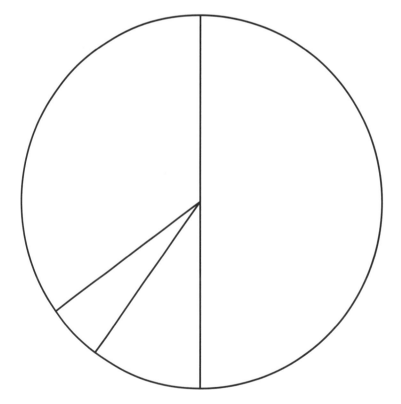

Creating and using circle graphs

Three hundred teenagers were asked what was the most important thing they learned from doing volunteer work. There were 75 who said to be kind and helpful; 105 said they learned to get along with others; 15 said they learned job skills; 60 said they learned to be a better person; and 45 said how to take care of children.

Here's how to make a circle graph to show the information above.

Step 1: Make a table to organize the information. Find the fraction and percent of the whole that each part of the information represents. The first one is done for you.

	Number	Fraction of total	Percent of total
Be kind and helpful	75	$\frac{75}{300} = \frac{25}{100}$	25%
Get along well with others			
Learn job skills			
Be a better person			
Take care of children			

Step 2: Find the angle measure that will show each part of the circle. Remember a circle has 360°. Multiply 360° by the fraction or percentage that the part represents.

Example: Be kind and helpful:
$\frac{25}{100}$ x 360° or 0.25 x 360° = 90°.
The angle measurement is 90°.

Step 3: Draw a circle and radius. Use a protractor to draw the angles. Label the sections with their percentages. Give the graph a title that tells what it is about.

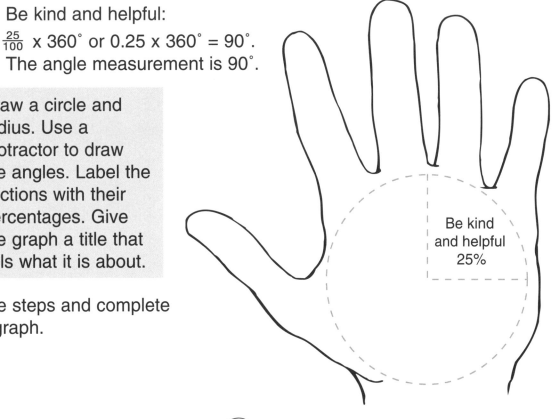

Refer to the steps and complete this circle graph.

Work with a partner. Create a circle graph about your favorite holiday or annual holidays in general.

First you must conduct a survey or research for information.

Describe your data and how you gathered it.

Draw the graph.

On Your Own—Circle graph

Line graphs show change over time.
This graph shows change in income from video rentals over a year's time.

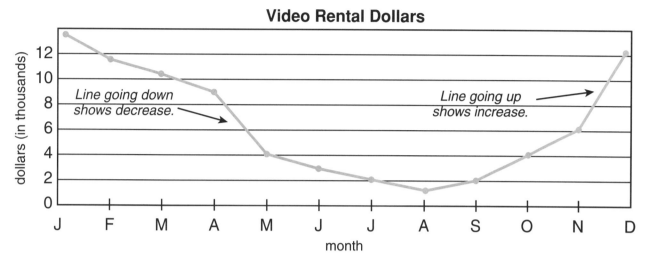

Example: To determine how much money was brought in from video rentals in November, find the point above November. Look in a straight line across to the left. $6000 worth of videos were rented.

Use the line graph to answer the questions.

1. How much money was brought in from video rentals in April? _____

2. What two months show $4000 in rentals? _____

3. During which month is the greatest number of videos rented? _____

4. During which month is the least number of videos rented? _____

5. Between which months is the greatest decrease in rentals? _____

   Between which months is the greatest increase in rentals? _____

6. What might be a reason for the increases and decreases in rentals?
   _____

7. Which month is your favorite for renting videos? _____

Reading and using line graphs

The line graph below shows how Juanita's bowling scores changed over 7 games. Use the table to complete the graph for games 8 and 9.

Game	1	2	3	4	5	6	7	8	9
Score	112	110	110	104	111	120	115	125	116

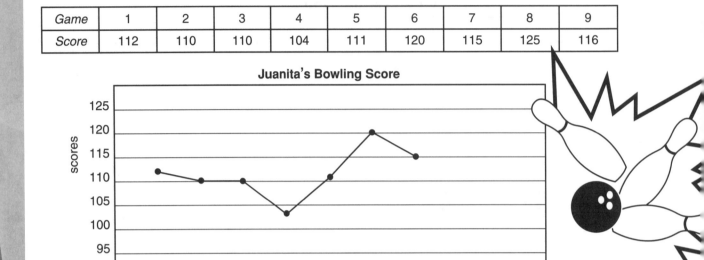

Use the line graph to answer the questions.

1. What was Juanita's highest score? _____

2. Between which two games did her score go down the most? _____

3. What happened between games 4 and 6? _____

4. Describe the line between the two games where there was no change in Juanita's score. _____

5. By how much did Juanita's scores vary? (Hint: Find the difference between the highest and lowest score.) _____

6. Juanita can say she usually scores above which number? _____

Completing and using line graphs

Here are the steps for making a line graph.

Step 1: Draw the horizontal axis and label it. Decide how much space to leave between each point.

Step 2: Determine a scale for the vertical axis. (Find the largest value. Determine a measurement to be equal to 1 unit. Will the highest point fit? If not, try 2 units for each and so on until you find a scale that will fit the space for the graph.) Draw the vertical axis and mark the scale. Draw the other two sides.

Step 3: Use the scale to mark the points. Connect the points.

Step 4: Give the graph a title that tells what the information is about.

This table shows the number of games won by the Wildcats each year.

Year	1993	1994	1995	1996	1997	1998	1999	2000
Games won	6	9	10	11	9	15	18	18

Use the data from the table above to make a line graph.

What can you say about the Wildcats since 1997?

_____

Making and using line graphs

Work with a partner. Each of you choose a stock you wish you could own. Track its performance over the last five years. If the company is not five years old, show its performance since it started.

Show what you learned on a line graph.

Write a paragraph about why you chose the stock that you did.

On Your Own—Line graph

A **double-line graph** shows two sets of related data on one graph so the two sets can be easily compared.
The graph below compares the average monthly temperatures in two cities.

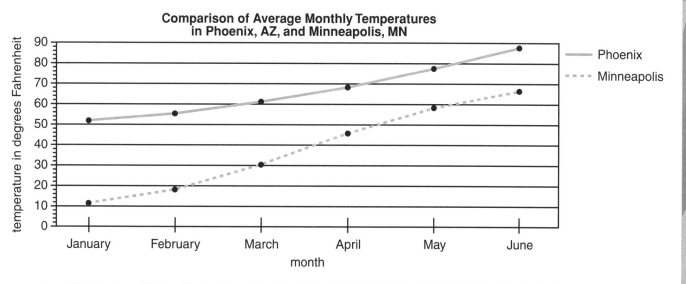

Example: To determine which city has the higher average temperature in April, first look at the key. The solid line represents temperatures in Phoenix. Find the temperatures for April on the graph. The solid line is higher, so Phoenix has the higher temperature.

Use the graph to answer these questions.

1. About what is the average monthly temperature in January for Minneapolis? _____

2. About what is the average monthly temperature for Phoenix in May? _____

3. In what month is there the greatest difference in temperature between the two cities? _____

4. During which two months is there the least difference in temperature between the cities? _____

5. How would you compare the change in temperature from winter to spring for the two cities? _____

6. Find the average monthly temperatures for January to June of the city nearest you. Plot the temperatures on the graph above.

7. On a separate sheet of paper, describe how the temperatures you graphed compare to the temperatures in Phoenix and Minneapolis.

Reading and using double-line graphs

Work with a partner. Conduct a survey to determine how much time each day you and at least 10 of your friends spend on homework and how much time is spent watching TV and/or playing computer games.

Show your findings on a double-line graph.

See if you can find what the national average is and write a paragraph about how you compare.

- Use bar graphs and pictographs to compare data.
- Use line graphs to show change over time.
- Use circle graphs or divided-bar graphs to show information as part of the whole.

Make a graph or display to show the data.
(Round numbers to the nearest 10 thousand.)

	Bushels
Grannie's Orchards	184,706
Big Red Farms	109,420
McAcres	154,918
Golden Valley Farms	198,783

Bushels of Apples Grown

Decide on the best way to show the heartbeat data. Then graph the data.

Heartbeats During Exercise

minutes	0	1	2	3	4	5	6	7	8	9	10
heartbeats	65	80	100	110	120	123	124	125	110	110	100

Selecting and making graphs

Make a graph or display to show the data.

Average Lengths of Snakes

King Cobra	Boa Constrictor	Anaconda	Python	Indigo	Indian Cobra
216 in.	144 in.	360 in.	300 in.	94 in.	72 in.

Make a graph or display to show the data.

Contents of York Landfill

Paper 38%	Glass 2%	Metal 14%	Yard Waste 11%	Food 4%	Plastic 18%	Other 13%

Choose any topic, conduct research, and create a graph of any type to display your findings.

Write a sentence or two explaining why you chose this type of graph.

Often the actual number of items being presented on a graph fall in between two scales. You can use that information to estimate quantities.

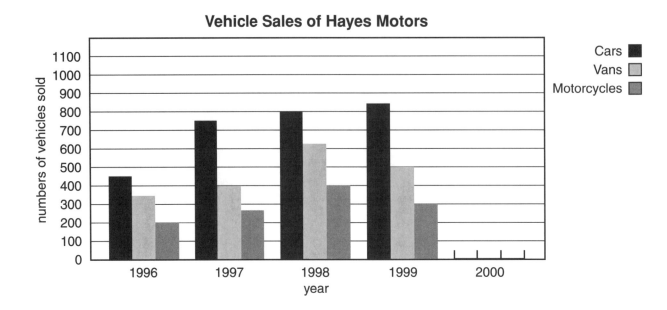

Example: To determine about how many cars were sold during 1997, look at the top of the bar for 1997. Look in a straight line to the scale. The top of the bar is about $\frac{2}{3}$ of the way between 700 and 800, or about 766. About 766 cars were sold in 1997.

1. Hayes Motors sold 1048 cars, 374 vans, and 322 motorcycles in 2000. Complete the graph.

Use the information on the graph to answer the questions.

2. About how many of each were sold in 1996?

   cars _____   vans _____   motorcycles _____

3. About how many more cars were sold in 2000 than in 1998? _____

4. In which year did Hayes Motors sell about 625 vans? _____

5. In which year was the number of motorcycles sold about 100 fewer than the previous year? _____

6. What was the approximate total number of vehicles sold in 1999? _____

7. What was the approximate change in the total number of vehicles sold in 1998 and in 1999? _____

Making and using graphs to estimate

Many graphs are good visual representations of patterns or trends, so past information can be used in predictions.

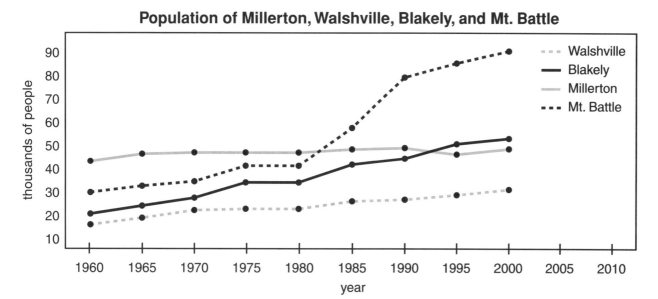

Example: To predict what the population in Walshville will be in the year 2005, look at the increase over the years shown. The population increases by about 2 thousand every 5 years, so you could predict that it would be about 30 thousand.

Use the graph to answer the questions.

1. What do you predict the population to be in the year 2005 for
   a. Blakely _____   b. Millerton _____
   c. Mt. Battle _____

2. Which of these regions has the greatest rate of increase? _____

3. Which of these regions do you predict will have the greatest population in the year 2005? _____

4. Lincoln had a population of 124,000 in 1990. What would you have to do to place it on the graph?
   _____
   _____

5. Is the population increasing or decreasing where you live? _____
   What do you think will happen in the next 5 years? _____
   _____

Making and using graphs to predict

How information is presented on a graph can affect the readers' opinions. For example, both of these graphs show the same information.

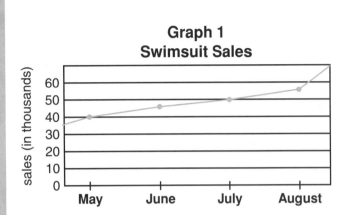

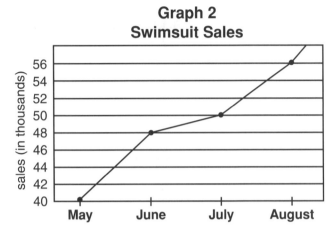

Use the information from the graphs to fill in the answers.

1. Swimsuit sales increased about 16 thousand over four months. Which graph makes 16 thousand look like a big increase? _____

2. What does a unit on the scale represent in Graph 1? _____
   Graph 2? _____

Making the units in the scale smaller makes the change look greater. This may be misleading.

Both graphs below show the number of boats sold this week at three dealers.

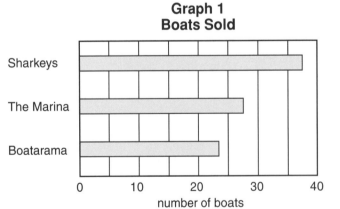

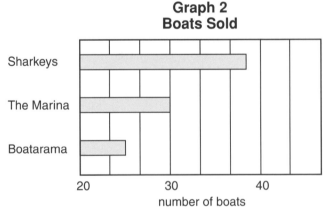

3. Which graph makes it appear that there is a great difference in sales between boat dealers? _____

When a graph does not begin at 0, it may be misleading.

4. What is the beginning number of boats sold on Graph 2? _____

Interpreting and evaluating graphs

Most newspapers and magazines include charts and graphs as part of presenting a story. The newspaper *USA Today* always includes graphs.

Collect at least ten graphs of a variety of types. On a separate sheet of paper, list each graph's title, type, and where you found it. If you can, include a copy of each graph with this lesson.

Now look at the graphs to see if there is one that you could use to estimate or predict. Describe it and tell why you chose it.

Could any graph be interpreted as having information that could be misleading? Explain your decision.

On Your Own—Using and evaluating graphs

You can use the graph at the right to find the *mean*, *median*, *mode*, and *range* of Amy's scores.

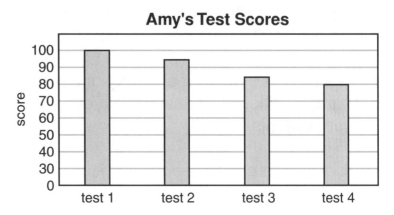

The **mean** is the average test score.
- Add the test scores. $\longrightarrow$
- Divide by the number of scores. $\longrightarrow$

$$\frac{100 + 95 + 85 + 80}{4} = \frac{360}{4} = 90 \text{ (mean score)}$$

The **median** is the middle number in a set of data that is arranged in order. In other words, there are the same number of entries above the median number as there are below it. The test scores arranged in order from greatest to least are 100, 95, 85, and 80. Since there is an even number of scores, the middle number would be halfway between 95 and 85. To find that number:

- Add the numbers. $\longrightarrow$
- Divide by the number of numbers. $\longrightarrow$

$$\frac{95 + 85}{2} = \frac{180}{2} = 90 \text{ (median score)}$$

The **mode** is the number that occurs most often in the set of data. In this set of data there is no number that occurs more often than the others. There is no mode. (There can be one mode, no mode, or several modes in a set of data.)

The **range** is the difference between the greatest and the least number in the set of data. 100 is the greatest score and 80 is the least.

$$100 - 80 = 20 \text{ (range)}$$

Find the mean, median, mode, and range for each set of data.

1. **Diving Scores**

   8   8   9   10   7   8   9

   mean: $\dfrac{8 + 8 + 9 + 10 + 7 + 8 + 9}{7} =$

   median: 10, 9, 9, 8, 8, 8, 7 _____
   mode: _____
   range: 10–7 = _____

2. **Baseball Scores**

   2   5   5   3   7   4   3   3

   mean: _____
   median: _____
   mode: _____
   range: _____

Determining mean, median, mode, and range

A **scattergram** is a graph that shows the relationships between two quantities. This graph shows the relationship of the time students studied and the test scores they received.

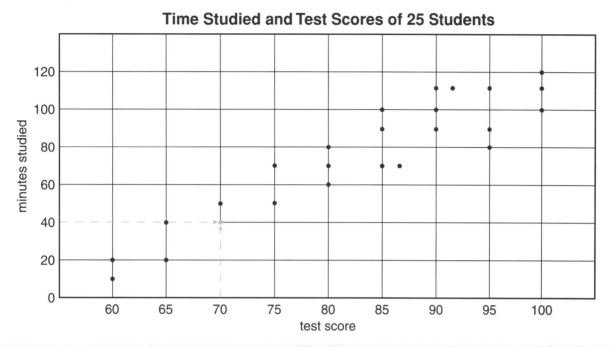

Example: To place the data on a scattergram for a person who studied 40 minutes and scored a 70 on the test, find 70, go up, and place a dot on 40 on the vertical scale.

1. Plot these scores and study hours on the graph.

Student	Minutes Studied	Score
J. S.	80	85
E. N.	90	100
C. M.	95	100
A. L.	85	90
K. H.	30	75

Use the graph to answer these questions.

2. How many students had a score of 85? _____

3. Did the 4 people with a score of 90 study the same number of minutes?

   _____

4. Does it appear that test scores increase or decrease as amount of time studied increases? _____

5. How would you describe the relationship between test score and time studied?

   _____

   _____

Completing and using scattergrams

You can organize data in a **stem and leaf plot**.

Length in Inches of Fish Caught
18 31 23 43 38 19 52 21 37 20 32 25 36 39 28 34 28

- Use the tens digits as *stems*.
- Fill in the *leaves* by writing the ones digit next to its stem.
- Arrange the leaves in order.

```
1 |                   1 | 8, 9                 1 | 8, 9
2 |                   2 | 3, 1, 0, 5, 8, 8     2 | 0, 1, 3, 5, 8, 8
3 |                   3 | 1, 8, 7, 2, 6, 9, 4  3 | 1, 2, 4, 6, 7, 8, 9
4 |                   4 | 3                    4 | 3
5 |                   5 | 2                    5 | 2
stems              stems   leaves           stems   leaves
```

1. What size fish was caught most often (the mode)? _____

2. What is the size of the fish in the middle of the set of data (median)? _____

3. What is the range of the sizes of the fish caught? _____

Make a stem and leaf plot for this data.

Number of fish caught today by members of the Hook, Line, and Sinker Club
28 31 40 52 21 41 54 27 35 43 21 60 36 29 37

stems | leaves

What is the mode of the data? _____ the median? _____
the range? _____

Completing and using stem and leaf plots

A **line plot** can be used to organize data.
Twenty students rated a new song, using a scale of 1 (worst) to 10 (best).

Ratings for New Song

9 8 10 5 6 9 7 8 9 10   6 10 7 9 8 10 9 5 7 10
  These ten have been plotted below.

A line segment numbered from 1 to 10 is drawn to represent the rating scale. Mark an X for each person's rating. (For example, three X's above a rating mean 3 people gave the song that rating.)

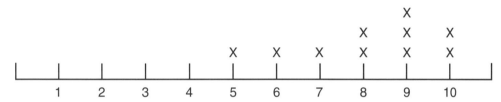

Complete the line plot.
Do you think the song will be a hit? _____ Explain. _____
_____

A **box plot** can be used to show a summary of data.
The weekly earnings of 13 students are shown in order from greatest to least.

A box is drawn around the middle half of the earnings. Whiskers are drawn from the ends of the box to the outer points. (The longer the box or whisker, the more spread out the data.)

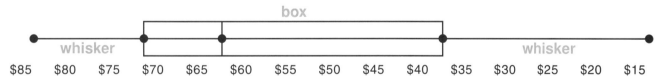

1. The middle half of the students surveyed have earnings between what two amounts? _____

2. Between what amounts are the earnings of the upper half? _____

3. Between what amounts are the earnings of the lower quartile? _____

Using line and box plots

A **histogram** is a bar graph that deals with frequency.

Sam rolled two dice 100 times and made a tally chart of the sum of the dice's numbers. Here is a partial histogram of his frequencies.

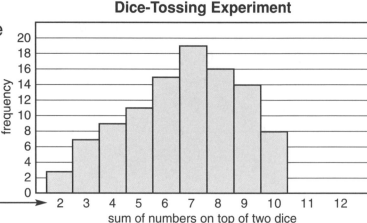

Example: Since the sum 2 occurred 3 times, he made a bar 3 units high above the sum of 2.

1. Complete the histogram to show that the sum of 11 came up 5 times and the sum of 12 came up 2 times.

2. Which sum from 2 to 12 came up most often? _____

A **frequency polygon** is a line graph that shows frequencies. Both ends of the graph are connected to the horizontal axis.

Here is a frequency polygon for the dice-tossing experiment shown above.

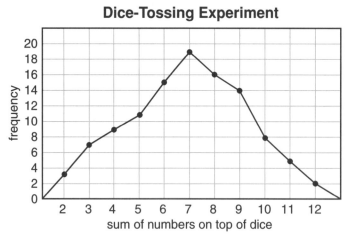

1. Suppose you were to play a game where you get a point for certain sums that come up when two dice are tossed. You can choose three sums for which you will earn a point. What sums would you choose? _____

   Why? _____

   _____

Using histograms and frequency polygons

Ask at least 50 people to guess the month of your birth. Then make a histogram or frequency polygon to show how frequently each month was chosen.

Design an experiment using coins that will demonstrate the concept of frequency. Describe and show the results.

Try this experiment. Toss two dice 108 times and record in the table below the sum of the two numbers that come up.

Sum	Tally	Frequency
2		
3		
4		
5		
6		
7		
8		
9		
10		
11		
12		

The **experimental probability** that an outcome will occur is the number of times an outcome occurred out of the total number of outcomes.

The experimental probability that the sum of 2 will occur when tossing two dice is

$\dfrac{\boxed{}}{108}$ ← $\dfrac{\text{number of times the sum of 2 occurred}}{\text{total number of times dice were tossed}}$ → is close to $\dfrac{\boxed{}}{36}$ expressed in lowest terms.

Use the results from the dice-tossing experiment above to estimate (in lowest terms) the experimental probability for each sum:

2	$\dfrac{\boxed{}}{108} \to \dfrac{\boxed{}}{36}$	3	$\dfrac{\boxed{}}{108} \to \dfrac{\boxed{}}{36}$	4	$\dfrac{\boxed{}}{108} \to \dfrac{\boxed{}}{36}$
5	$\dfrac{\boxed{}}{108} \to \dfrac{\boxed{}}{36}$	6	$\dfrac{\boxed{}}{108} \to \dfrac{\boxed{}}{36}$	7	$\dfrac{\boxed{}}{108} \to \dfrac{\boxed{}}{36}$
8	$\dfrac{\boxed{}}{108} \to \dfrac{\boxed{}}{36}$	9	$\dfrac{\boxed{}}{108} \to \dfrac{\boxed{}}{36}$	10	$\dfrac{\boxed{}}{108} \to \dfrac{\boxed{}}{36}$
11	$\dfrac{\boxed{}}{108} \to \dfrac{\boxed{}}{36}$	12	$\dfrac{\boxed{}}{108} \to \dfrac{\boxed{}}{36}$		

1. Which three sums occurred more times than all the others? _____

2. Which two sums occurred fewer times than all the others? _____

Experimenting and using experimental probability

You can determine the **theoretical probability** of an event.
Here is a chart of all possible outcomes or ways the sum of two dice can come up.

Sum	Sum	Sum	Sum	Sum	Sum
2	3	4	5	6	7
3	4	5	6	7	8
4	5	6	7	8	9
5	6	7	8	9	10
6	7	8	9	10	11
7	8	9	10	11	12

To find the probability that the sum of 6 will occur, you can use this formula:

$$P(\text{Event}) = \frac{\text{number of favorable outcomes}}{\text{number of possible outcomes}}$$

The probability of a sum of 6 coming up:

$$P(6) = \frac{\text{number of ways a sum of six can come up}}{\text{number of possible ways}} \rightarrow \frac{5}{36} \rightarrow$$ You can predict that a sum of 6 will come up 5 out of 36 times.

1. You can use the chart above to find the probability of a sum coming up on the dice. Complete each exercise.

   P(2) = ___/36   P(3) = ___/36   P(4) = ___/36   P(5) = ___/36   P(7) = ___/36

   P(8) = ___/36   P(9) = ___/36   P(10) = ___/36   P(11) = ___/36   P(12) = ___/36

2. How do the answers above compare to the experimental probabilities you found on page 52? _____

   _____

3. What is the theoretical probability that a head will come up when you toss a coin? P(H) = _____

4. What is the theoretical probability that this spinner will land on a 5? P(5) = _____

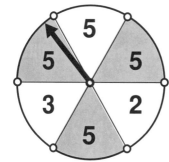

5. What is the theoretical probability of picking an M?

   P(M) = _____

Using theoretical probability

One event that has no effect on another event is called an **independent event**. If you spin the spinner and then pick one of the cards, the events are independent of one another. Where the spinner stops has no effect on which card is drawn.

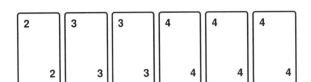

If two events are independent, the probability that both will occur can be found by multiplying the individual probabilities.

Here's how to find the probability of the spinner landing on blue and drawing a 3 card.

$$P(\text{blue}, 3) = P(\text{blue}) \times P(3)$$
$$\frac{3}{8} \times \frac{2}{6}$$
$$\frac{3}{8} \times \frac{1}{3} = \frac{3}{24} = \frac{1}{8}$$

> You can predict that the spinner will land on blue and a card with 3 on it will be drawn 1 out of 8 times.

Find the probability for spinning the spinner and drawing a card.

1. P (yellow, 4) _____
2. P (green, 2) _____
3. P (blue, 4) _____
4. P (blue or yellow, 3) _____
5. P (green, 2 or 3) _____
6. P (green or yellow, number greater than 2) _____

Suppose you roll a die and draw one of these cards.

Find the probability.

7. P (1, ●) _____
8. P (3, ▲) _____
9. P (6, ■) _____
10. P (number less than 4, ▲) _____
11. P (multiple of 2, ●) _____
12. P (not 6, ■) _____

Determining probability of independent events

When one event does affect another event, they are **dependent events.**

If prizes for a party are wrapped in the same size box and are sitting in a row on the table, what is the probability of the first and second prizes chosen both being games?

The first event or choice does affect the second event. After one of the 9 prizes is chosen, there is one fewer prize. So the events are dependent.

To find the probability of dependent events, multiply.

P (games, games) = $\frac{4}{9} \times \frac{3}{8} = \frac{12}{72} = \frac{1}{6}$

4 out of the 9 boxes have games, so the chance of getting a game on the first try is $\frac{4}{9}$.

Only 8 prizes are left for the second draw.

You can predict 1 out of 6 times that the first two prizes drawn will be games.

Find the probability of choosing the prizes shown above.

1. P (game, key chain) $\frac{\Box}{9} \times \frac{\Box}{8} =$

2. P (key chain, pen)

3. P (pen, game)

4. P (key chain, key chain)

5. P (pen, pen)

6. P (game, pen)

7. P (pen, key chain)

8. P (key chain, game)

Determining probability of dependent events

The **odds** in favor of an event is the **ratio** of the number of favorable outcomes to the unfavorable outcomes.

Candy bars of the same size are in a bag.

Example: What are the odds in favor of choosing a chocolate bar?

There are 4 chocolate bars (favorable outcomes).

There are 6 bars other than chocolate (unfavorable outcomes).

The odds in favor of choosing a chocolate bar out of the bag without looking are

$\dfrac{4}{6}$ ← number of favorable outcomes
← number of unfavorable outcomes

Read: 4 to 6

The odds of choosing a chocolate bar are 4 to 6.

Notice that the sum of two numbers in the ratio equals the total number of things. There are 10 candy bars in all.

Find the odds of choosing each type of candy bar.

1. nut bar _____   2. caramel _____   3. strawberry _____

A cookie jar contains 6 chocolate chip cookies, 4 oatmeal cookies, and 3 peanut butter cookies.

If you choose a cookie without looking, what are the odds in favor of choosing

4. chocolate chip? _____   5. oatmeal? _____   6. peanut butter? _____

There are 12 cupcakes in a box—5 chocolate, 4 vanilla, and 3 spice.

If you choose a cupcake without looking, what are the odds in favor of choosing

7. chocolate? _____   8. vanilla? _____   9. spice? _____

Determining odds

You can use a **tree diagram** to find all the possible outcomes or combinations.

Claire bought three T-shirts and two pairs of shorts. How many different outfits can she wear?

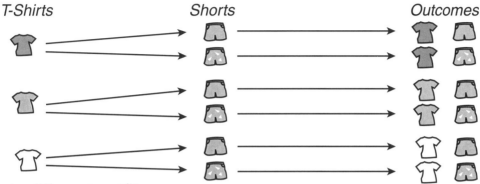

There are six different outfits.

You can also use the **counting principle** to find the number of all the possible outfits or outcomes.

number of choices for T-shirts		number of choices for shorts	
3	X	2	= 6 possible outfits

1. Complete the tree diagram to find all possible combinations of outfits.

   Pants: navy, black, gray   Shirts: blue, yellow, white, red

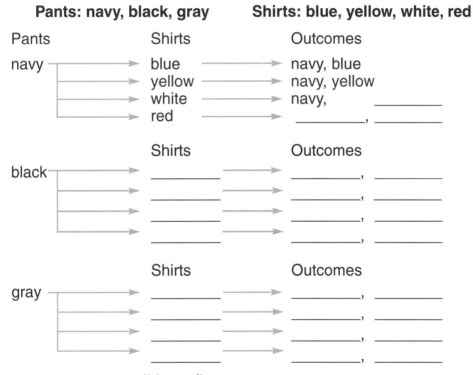

   There are _____ possible outfits.

2. Use the counting principle to find the number of all the possible combinations of outfits.

   _____ X _____ = _____

Completing and using tree diagrams

Each possible arrangement or combination in a definite order is a **permutation**.

There are 5 dogs in a class at a dog show.

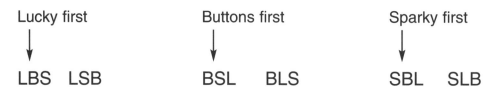

Lucky   Dusty   Muffin   Buttons   Sparky

Ribbons will be awarded for first, second, and third place. Here are the ten possible combinations of three "top dogs." (Combinations are given using each dog's initial.)

LDM   LDB   LDS   LMB   LMS   LBS   DMB   DMS   MBS   BSD

Suppose Lucky, Buttons, and Sparky are the three "top dogs." Because they are awarding 1st, 2nd, and 3rd places, the arrangement of the dogs in *order* is important. The six arrangements or permutations in 1st, 2nd, and 3rd place order of Lucky, Buttons, and Sparky are

Lucky first → LBS   LSB

Buttons first → BSL   BLS

Sparky first → SBL   SLB

Suppose prizes for 1st, 2nd, 3rd, and 4th place are awarded in the class at the dog show.

What are the five combinations of 4 "top dogs"?

_____   _____   _____   _____   _____

Suppose Lucky, Buttons, Muffin, and Sparky are the 4 "top dogs." Complete the table to show the possible permutations or arrangements of dogs in 1st, 2nd, 3rd, and 4th places. (A few examples are provided.)

Lucky finishes 1st	Buttons finishes 1st	Muffin finishes 1st	Sparky finishes 1st
L B M S	B _ _ _	M _ _ _	S _ _ _
L B S M	B _ _ _	M _ _ _	S _ _ _
L M B S	B _ _ _	M _ _ _	S _ _ _
L M S B	B _ _ _	M _ _ _	S _ _ _
L S B _	B _ _ _	M _ _ _	S _ _ _
L S M _	B _ _ _	M _ _ _	S _ _ _

How many permutations are there? _____

Determining permutations

Try this experiment. Have someone put 5 marbles or buttons in a bag so you do not know how many of each color or what colors are in the bag. Without looking, draw a marble out of the bag, record the color in the tally chart, and put the marble back in the bag. Do this 100 times. Then complete the chart.

Color	Tally	Experimental Probability	Estimate, or express as a simple fraction with denominator of 5
		$\overline{100}$	$\overline{5}$

You discovered earlier that actual and experimental probability should be close.

1. Suppose there are 2 blue marbles in the bag. The actual probability would be $\frac{2}{5}$. What would the experimental probability be close to? _____

2. Look in your table. About how many out of 5 times did you draw the first color? _____

3. Based on the results of your experiment, how many of each color do you think are in the bag? _____
   For example, if a red marble has an experimental probability of about $\frac{4}{5}$, then you can predict that there are 4 red marbles in the bag.

Check the bag to see if your predictions are correct.

4. Suppose the actual probability of drawing a white marble is $\frac{1}{4}$. About how many times out of 100 would you expect to draw a white marble? _____
   If there are 8 marbles in the bag, how many are white? _____

Determining experimental and actual probability

Answer Key

Page numbers are indicated in color.

4
Answers will vary.

5
1. 5
2. 3
3. less than 1
4. 1,540,000
5. 39,610,000
6. 55,060,000

6
1. Subtotal: $ 8.40
   Tax: $ 0.50
   Total: $ 8.90
2. Subtotal: $ 9.50
   Tax: $ 0.57
   Total: $10.07
3. Subtotal: $ 7.75
   Tax: $ 0.47
   Total: $ 8.22
4. Subtotal: $ 8.15
   Tax: $ 0.49
   Total: $ 8.64

7
1. $10.00
2. 2 pizzas $8.00
3. Yes
4. $6.85
5. $2.88
6. $0.50
7. $82.50

8
1. 22 hours $115.50
2. 25 hours $168.75
3. 30 hours $165.00
4. 32 hours $192.00
5. 36 hours $219.60
6. 35 hours $222.25
7. 37 hours $216.45

no. of hours	tally	frequency
3	I	1
4	IIII	4
5	ⅢⅡ	5
6	ⅢⅡ ⅢⅡ	10
7	ⅢⅡ II	7
8	ⅢⅡ III	8

1. 4 hours
2. 6 hours
   3 hours

9
1. 11 miles
2. $1.00
3. $11.00
4. $4.25
5. $1.40
6. $6.00
7. $2.05
8. $41.00
9. 117 miles
10. $239.85

10
1. 49 min.   49 min.
2. 40 min.
3. 48 mph   80 mph
4. $8.00 $2.20
   $4.50 $2.75

11
Answers will vary.

12
1. $1098.00
2. $118.60
3. $1532.70
4. $637.90
5. $396.80

13

Appliance	WATTS USED					
	1 hr.	2 hr.	3 hr.	6 hr.	8 hr.	24 hr.
Hair Dryer	1600	3200	4800	9600	12,800	38,400
Stereo	150	300	450	900	1200	3600
Refrigerator	750	1500	2250	4500	6000	18,000
Electric Stove	12,000	24,000	36,000	72,000	96,000	288,000
Color TV	200	400	600	1200	1600	4800
Computer	300	600	900	1800	2400	7200
100-Watt Light Bulb	100	200	300	600	800	2400

1. $0.14
2. $1.44
3. $2.88
4. $0.05
5. $0.05
6. less than a penny
7. $0.03

14
1. $8.65 $103.80
2. $23.54 $564.96
3. $1195.92 $195.92
4. $1881.80

15

Item	Catalog Number	How many	Price for 1	Total Price	Shipping Weight lb
Radio	416-5342	2	$29.99	$59.98	6.2
Scooter	614-58310	1	$49.99	$49.99	9.2
			Total	$109.97	15.4
			Tax	$6.59	
			Shipping and Handling	$5.55	
			Total Cost	$122.11	

16
Answers will vary.

17
1. $0.50
2. 800 shares
3. $9.00
4. $48 $\frac{7}{8}$ = $48.875
5. $1425
6. $7650
   $0.75 per share or $300

18
Answers will vary.

19
1. how long whales are
2. Blue Whale
3. Killer Whale
4. Fin Whale
5. 2 meters
6. 3 meters
7. 21 meters

20
1. 1380 feet — table
2. Sears Tower — graph
3. Kuala Lumpur, Maylasia — table
4. Petronas Towers — graph
5. 233 feet — table
6. 3 buildings — either

21
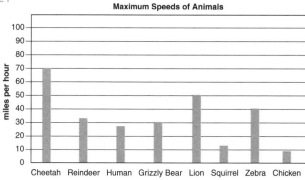
1. No
2. Yes

22
Answers will vary.

23
1.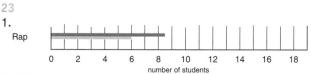
2. 5 boys
3. 6 girls
4. boys
5. country
6. girls: rock, pop, country/rap/hip hop
   boys: rock, rap, country, pop, hip hop
7. 76 students

24
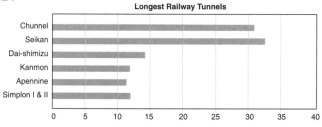
1. sons
2. Graphs will vary.

25
1. a. 5,500,000 b. 4,000,000
   c. 3,000,000 d. 2,500,000
2. 1,000,000
3. 1950
4. decreased by 3,300,000
5. $1\frac{1}{2}$ barns
6. 5 barns

26
1. 4 bicycles complete the graph
2. Multiply the number of pictures drawn by 50. Check that it equals 200.
3. Answers will vary.
4. 25 bicycles
5. Wednesday
6. 175 bicycles
7. $3\frac{1}{2}$ $\frac{3}{4}$
8. 950 bicycles

27

U.S. Population in Millions

year	1950	1960	1970	1980	1990	2000
number of people	152.3	180.7	205.1	227.2	249.5	281.4
	150	180	210	230	250	280

U.S. Population in Millions

(pictograph, $\text{\Large\textdagger}$ = 20 million people)

28
Answers will vary.

29
1. blue
2. red
3. blue and red
4. 8%
5. $\frac{1}{4}$
6. 24 people
7. 90 people

30
Bookworms' Book Sale

Total Books Sold: 80			
Type of Book	Number Sold	Fraction of all Books Sold	Percent Sold
Mysteries	40	$\frac{40}{80} = \frac{1}{2}$	$\frac{1}{2} = \frac{50}{100} = 50\%$
Novels	28	$\frac{28}{80} = \frac{7}{20}$	$\frac{7}{20} = \frac{35}{100} = 35\%$
Reference	8	$\frac{8}{80} = \frac{1}{10}$	$\frac{1}{10} = 10\%$
Other	4	$\frac{4}{80} = \frac{1}{20}$	$\frac{1}{20} = 5\%$

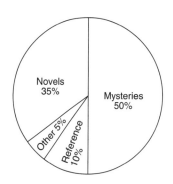

31

	Number	Fraction of total	Percent of total
Be kind and helpful	75	$\frac{75}{300} = \frac{25}{100}$	25%
Get along well with others	105	$\frac{105}{300} = \frac{7}{20}$	35%
Job skills	15	$\frac{15}{300} = \frac{1}{20}$	5%
Be a better person	60	$\frac{60}{300} = \frac{1}{5}$	20%
Take care of children	45	$\frac{45}{300} = \frac{9}{60} = \frac{3}{20}$	15%

32
Answers will vary.

33
1. $9000
2. May and October
3. January
4. August
5. April and May
   November and December
6. Possible answer: In warmer weather people may spend more time doing outdoor activities.
7. Answers will vary.

34
1. 125
2. Games 8 and 9
3. Her score improved by 16 pins.
4. The line is exactly horizontal or level.
5. 21 pins
6. 110

35

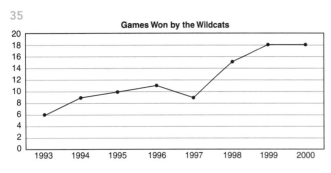

They have been winning more games. The team has improved.

36
Answers will vary.

37
1. About 11°
2. About 78°
3. January
4. May and June
5. The temperature rises in both cities, but Minneapolis has a greater change.
6. Answers will vary.
7. Answers will vary.

38
Answers will vary.

39
Answers to both exercises will vary. We show answers as a bar and a line graph, respectively.

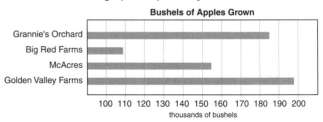

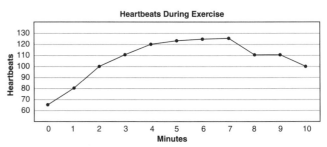

40
Answers may vary by type of graph. We show bar and circle graphs, respectively.

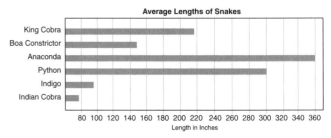

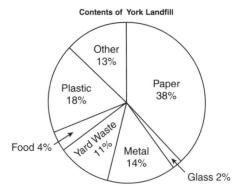

41
Answers will vary.

42
1.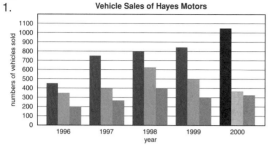
2. cars: about 450 vans: about 340 motorcycles: about 200
3. 250 cars
4. 1998
5. 1999
6. 1650 vehicles
7. It decreased about 150 vehicles.

43
Answers will vary slightly.
1. a. 60,000
   b. 50,000
   c. 95,000
2. Mt. Battle
3. Mt. Battle
4. Increase or extend the scale. Put Lincoln in the key.
5. Answers will vary.

44
1. Graph 2
2. 10,000 swimsuits
   2000 swimsuits
3. Graph 2
4. 20 boats

45
Answers will vary.

46
1. 8.4   2. 8
   8.5   4.5
   8   3
   3   5

47
1. Check students' graphs.
2. 4 students
3. No
4. increase
5. the more time spent studying, the higher the test score

48
1. 28 inches
2. 31 inches
3. 34 inches

stems	leaves
2	1, 1, 7, 8, 9
3	1, 5, 6, 7
4	0, 1, 3
5	2, 4
6	0

21 fish
36 fish
39 fish

49
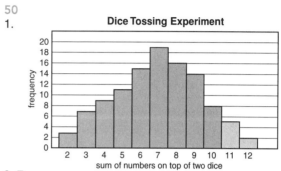

Yes. Most people gave the song a rating of 8, 9, or 10.

1. $71.50 and $39
2. $64 to $84
3. $39 to $15

50
1. [Dice Tossing Experiment histogram]
2. 7

1. 6, 7, and 8. They come up most often, so you would get the most points.

51
Answers will vary.

52
Answers will vary. However, 7 should come up the most, 6 and 8 next, then 5 and 9, 4 and 10, 3 and 11, and 2 and 12.

1. 6, 7, 8
2. 2 and 12

53
1. $\frac{1}{36}, \frac{2}{36}, \frac{3}{36}, \frac{4}{36}, \frac{6}{36}$
   $\frac{5}{36}, \frac{4}{36}, \frac{3}{36}, \frac{2}{36}, \frac{1}{36}$
2. Answers will vary.
3. $\frac{1}{2}$
4. $\frac{4}{6}$ or $\frac{2}{3}$
5. $\frac{2}{11}$

54
1. $\frac{4}{8} \times \frac{3}{6} = \frac{12}{48} = \frac{1}{4}$ 2. $\frac{1}{8} \times \frac{1}{6} = \frac{1}{48}$
3. $\frac{3}{8} \times \frac{3}{6} = \frac{9}{48} = \frac{3}{16}$ 4. $\frac{7}{8} \times \frac{1}{3} = \frac{7}{24}$
5. $\frac{1}{8} \times \frac{3}{6} = \frac{3}{48} = \frac{1}{16}$ 6. $\frac{5}{8} \times \frac{5}{6} = \frac{25}{48}$
7. $\frac{1}{6} \times \frac{3}{6} = \frac{3}{36} = \frac{1}{12}$ 8. $\frac{1}{6} \times \frac{2}{6} = \frac{2}{36} = \frac{1}{18}$
9. $\frac{1}{6} \times \frac{1}{6} = \frac{1}{36}$ 10. $\frac{3}{6} \times \frac{2}{6} = \frac{6}{36} = \frac{1}{6}$
11. $\frac{3}{6} \times \frac{3}{6} = \frac{9}{36} = \frac{1}{4}$ 12. $\frac{5}{6} \times \frac{1}{6} = \frac{5}{36}$

55
1. $\frac{4}{9} \times \frac{3}{8} = \frac{12}{72} = \frac{1}{6}$ 2. $\frac{3}{9} \times \frac{2}{8} = \frac{6}{72} = \frac{1}{12}$
3. $\frac{2}{9} \times \frac{4}{8} = \frac{8}{72} = \frac{1}{9}$ 4. $\frac{3}{9} \times \frac{2}{8} = \frac{6}{72} = \frac{1}{12}$
5. $\frac{2}{9} \times \frac{1}{8} = \frac{2}{72} = \frac{1}{36}$ 6. $\frac{4}{9} \times \frac{2}{8} = \frac{8}{72} = \frac{1}{9}$
7. $\frac{2}{9} \times \frac{3}{8} = \frac{6}{72} = \frac{1}{12}$ 8. $\frac{3}{9} \times \frac{4}{8} = \frac{12}{72} = \frac{1}{6}$

56
1. $\frac{3}{7}$ 2. $\frac{2}{8}$ 3. $\frac{1}{9}$
4. $\frac{6}{7}$ 5. $\frac{4}{9}$ 6. $\frac{3}{10}$
7. $\frac{5}{7}$ 8. $\frac{4}{8}$ 9. $\frac{3}{9}$

57
1.
Shirts	Outcomes
white	navy, white
red	navy, red
blue	black, blue
yellow	black, yellow
white	black, white
red	black, red
blue	gray, blue
yellow	gray, yellow
white	gray, white
red	gray, red

12

2. 3 x 4 = 12 possible outfits

58
LDMB, LMBS, LBSD, LSDM, DMBS
Note: The letters in each group may be in any order.

LBMS, BLMS, MLBS, SLBM
LBSM, BLSM, MLSB, SLMB
LMBS, BMLS, MBSL, SBLM
LMSB, BMSL, MBLS, SBML
LSBM, BSLM, MSLB, SMLB
LSMB, BSML, MSBL, SMBL
24

59
1. $\frac{2}{5}$
2. Answers will vary.
3. Answers will vary.
4. 25 times
   2 marbles